Whitfield Ward

Singers' Throat Troubles

their cause and cure - course of lectures delivered at the Grand

Conservatory of Music, during the season 1883-84

Whitfield Ward

Singers' Throat Troubles
their cause and cure - course of lectures delivered at the Grand Conservatory of Music, during the season 1883-84

ISBN/EAN: 9783337223618

Printed in Europe, USA, Canada, Australia, Japan

Cover: Foto ©berggeist007 / pixelio.de

More available books at **www.hansebooks.com**

Singers'

Throat Troubles

Their Cause and Cure

Course of Lectures delivered at the Grand Conservatory
of Music, during the season 1883–84.

BY

WHITFIELD WARD, A. M. M. D.

Physician to the Metropolitan Throat Hospital; author of
"The Throat in Relation to Singing," etc.

PUBLISHED BY
THE GRAND CONSERVATORY PUBLISHING CO.
46 West 23d Street, New York
1885

PRESS OF
WILLIAM S. GOTTSBERGER,
11 Murray St., New York.

PREFACE.

In this little work which is intended as a companion to "The Throat in Relation to Singing," I have endeavored to present a series of pictures of the most prominent ailments peculiar to the vocal organs of singers.

Believing that many voices are annually lost through ignorance and mismanagement, I have, for the express purpose of remedying this evil, undertaken in the following pages to give the vocalist an insight into many of these difficulties.

THE AUTHOR,

208 West 25th Street. New York City.

SINGERS' THROAT TROUBLES.

THEIR CAUSE AND CURE.

CHAPTER I.

The Causes of Throat Troubles and the Modern Methods of Treatment.

IN no other branch of the medical sciences have such rapid strides been made within the past decade as in Laryngology, or the diagnosis and treatment of diseases peculiar to the upper portion of the respiratory tract. Although this science had its beginning in the discovery of the principles of examining the throat by Garcia, it is only within a few years that the important ailments of the throat have been properly studied and the knowledge of these diseases made thorough and comprehensive.

Now, while the laity possesses a fair idea of the affections of many other portions of the human body, it is safe to say that they have very little knowledge of the throat, not only regarding its diseases, but also concerning its care.

This is especially noticeable in singers and singing teachers, many of the latter class exhibiting the greatest ignorance of ordinary points of vocal hygiene. Such ignorance on the part of singing teachers is entirely inexcusable and is engendered by the false idea that these things are without their profession and belong solely to the physician. This is a great mistake and productive of much unnecessary sickness. How can the master by giving such advice as shall prevent disease trespass upon the functions of the physician? Such an idea is preposterous. The vocal trainer by acquainting himself with the ordinary laws of vocal Hygiene, and they certainly are simple enough, can be the means of doing incalculable good to his pupil. Many singers suffer from serious throat troubles which could have been easily avoided by a little good advice at the outset of their difficulties.

The science of laryngology then dating from the adoption of the laryngoscope by the medical profession, has grown in a comparatively short space of time to wonderful proportions. By the aid of this truly marvelous instrument the hidden recesses of the vocal apparatus are brought to light, which enables the seeker after truth to study its internal workings. Many wonderful things have we learned through the agency of the little mirror of the laryngoscope which, when inserted into the mouth of the person under examination, gives upon its reflecting surface a perfect view of that wonderful little organ the human larynx. Affections which were hitherto unheard of, have been, since the employment of the laryngoscopic apparatus, discovered and their nature and cause thoroughly investigated. The true science of medicine lies chiefly in the diagnosing of disease and not, as is the popular belief, in the art of healing. When once an affection has been successfully diagnosed and its cause and peculiarities are thoroughly understood, its cure, provided it be a curable complaint, is simple enough. Nature supplying the necessary drugs and only requiring a proper selection which, as we all know, is based upon experience and experimentation.

If the laryngoscope has been of service in medical science, it has been none the less valuable in the art of voice-training. Although many self-constituted members of the musical profession have from time to time ridiculed the laryngoscope and its wonderful powers, stating that it is of no practical value to the singing teacher, yet the fact stands out in bold relief, that this instrument is of the utmost worth and importance to the much abused art of voice training. One gentleman in particular hailing from that centre of aesthetic art, Boston, took the trouble to assail this most valuable instrument, a short time since, in a five-column article contributed to a musical journal of this city. The subject matter contained in the paper referred to above, was of such a nonsensical character and displayed such gross ignorance on the part of the writer, that, although it was personally aimed at myself, I totally ignored it. By means of the laryngoscope, an apparatus which consists of three instruments, an illuminator supplying the light, a large mirror strapped to the forehead to cast the light into the throat, and a small mirror to insert into the mouth, we have been enabled to study the mechanism of those wonderful little bodies, the vocal cords, which, as we all know, are the prime factors in the production of musical tones. If the skeptic, be he master or pupil, could see as I have repeatedly, the beautiful movements which characterize the mechanical action of the vocal cords during vocalization, his skepticism would vanish as a cloud, and he would at once become an ardent advocate and admirer of the apparatus which enabled him to explore the laryngeal organ. The laryngoscope enables us to find defects in the vocal organism which would otherwise have remained undiscovered. These defects which are in some cases trivial, and in others great, will oftentimes, if neglected, result in a total loss of the vocal powers. The singer is often unaware of any trouble in the parts, he simply feels that his vocal abilities are not the same as they were. Now, if perchance he has a careful and judicious master, the latter personage will at once insist upon a laryngeal examination, which, if it is undergone, will at once reveal the difficulty and enable it to be remedied in a short time. But, if on the contrary, the sufferer has a non-believer in laryngoscopy for a teacher, this individual will make light of his complaints and will assure the pupil that there is nothing the matter with his vocal organs. What will be the result? Why the voice will be more and more impaired and after a varying period of time, which depends, of course, on the nature of the defect, the ability of the victim to vocalize will be seriously compromised if it is not entirely lost.

There is almost a universal dread among singers of undergoing any treatment whatsoever of the throat, consequently many through fear or ignorance suffer from and allow to become chronic affections, which if attended to in time would have required but the simplest and shortest treatment. The principal effect of this prejudice is the injury of the vocal organs which is sure to result and frequently become permanent. I am very sorry to say that this dread is fostered among pupils by teachers who invariably advise their wards to beware of "Throat doctors" informing them that not only will their vocal apparatus be injured, but that when once treatment has been begun they will have to continue it for an indefinite period. Such an assertion shows either the grossest ignorance or an inordinate love of gain. From my long and varied experience with some of these teachers, I am forced to state that in the majority of instances ignorance is not the cause of their giving such advice. How, you will immediately ask, can the teacher gain by advising against professional treatment of affected vocal organs? In two ways, when the throat is the seat of a slight inflammation, or, in other words, the afflicted person has taken cold, the very first thing the physician will advise is to stop all singing. Why? because when the larynx is inflamed vocalization only inhances the difficulty. In the treatment of inflammatory diseases, no matter in what portion of the human economy they may be located, the first principal is absolute rest of the parts. This precept is especially true of the larynx, because the vocal cords, the vibrating reeds producing sound when inflamed are greatly thickened, and all acts of vocalization cause a constant rubbing together of these important little bodies which, as can be readily seen, will increase the inflammatory action, exactly the same effect will be produced, as if an inflamed surface on one hand were constantly rubbed by the other, namely: A continuance of, and an increase in the diseased condition. Now when the physician advises the singer to stop, for the time being his studies, he stops the flow of ducats into the pockets of the unscrupulous teacher. Of course it is only when the inflammation is slight that the foregoing remarks are applicable. When the inflammatory

action is great the singer will be obliged to rest for the simple reason
that the congestive swelling of the vocal cords is so great that these
bodies will be unable to vibrate sufficiently to produce vocal sounds. This
is indeed a most fortunate thing for the sufferer, and it has been the
means of preserving many voices. When the inflammatory action is
slight, and the resultant hoarseness or huskiness correspondingly so,
then it is that the bad advice of a bad teacher will be productive of
considerable harm. I have been repeatedly told by singers, that their
teachers have advised them to sing off their colds, that although the voice
was hoarse exercise would remove the difficulty. In some instances it
will do this, but only for the time being, and at the expense of the
sufferer's voice. Why ? because it will call into play the large external
laryngeal muscles, bodies that ordinarily have nothing to do with vocalization, and whose action will in time be productive of the greatest
injury to the voice. The second answer to the question given above, is
that many masters are afraid to let an expert see the result of their labor.
The experienced examiner can, with the aid of the laryngoscope, tell in
an instant whether the voice has been properly trained or not. This
may seem a broad statement but it is true in every respect. When the
voice has been properly trained the vocal cords present a beautiful
appearance not only as regards color and conformity, but also as
regards their physiological action. Their color is a pearly white and
when approximated or adducted, they present the appearance of two
firm bands of tissue seperated by a fine and somewhat symmetrical
opening. This interval is known as the chink of the glottis, and is
always present during ordinary acts of vocalization. The function of
the chink of the glottis is to provide a passage for the exit of air which
is a prime factor in the production of sound. When, however, the
vocalist has been the victim of bad training, the vocal cords instead of
presenting the above appearance are flabby and relaxed and more or
less thickened and congested. The chink of the glottis is also partially
or totally obliterated while the entire larynx is somewhat relaxed and frequently congested. What does all this show ? simply that the methods
of teaching employed have been faulty. A relaxed condition of the
vocal cords is the condition most often resulting from improper tuition.
No pair of vocal cords can perform their duty properly unless they
possess a certain degree of firmness. If they are in any manner
relaxed it will be impossible for the delicate vocal muscles to tense
them to such a degree as to produce a tone properly. How then can
a vocalist with such a pair of cords sing ? By bringing to bear on the
cords the large external laryngeal muscles, bodies which ordinarily have
nothing to do with vocalization ; or in plainer language by forcing the
voice. Such action will most certainly be followed by loss of voice. I
have seen this result too often to admit of any doubt.

I have often warned students against those whom I know have harmed their voices. In doing this I simply perform my duty. I have no especial favorites among teachers but when I see a larynx which has been developed by a good and careful trainer I am only too willing to applaud his work.

Very many vocal masters are wont to defend themselves by asserting that the injury to their pupils voices was accomplished by some former instructor. All I can say in answer to this is that if the master after a certain period of time, finds that there is some physical obstacle to the proper training of a pupil's voice, he should advise him to have the vocal apparatus thoroughly examined with the laryngescope by an expert, and be guided by the examiner's diagnosis. If the teacher does not do this, but, on the contrary, continues his endeavors to train such a case, he renders himself partially culpable for his pupil's condition because his tuition increases the difficulty. It is a golden rule, and it should always be in the mind of the vocal-master, that unless the entire vocal apparatus is in a proper condition it will be absolutely impossible for the scholar to receive any good whatsoever from his instruction no matter how perfect may be the method employed. The affections of the throat met with in medical practice are not dissimilar from the diseases encountered in other regions of the body.

Inflammation occurs in its various grades, we likewise have the products of inflammation, such as swellings and tumors of various kinds and finally we meet with many nervous ailments. The mucus membrane with which the entire respiratory tract is lined is with but few exceptions the seat of throat disease. The reason of this peculiar proclivity is partly from the exposed position of the membrane lining the throat, and partly from extension of diseases existing in adjacent parts which in this portion of the body, are peculiarly prone to spread.

Most of the inflammatory affections of the throat commence in the pharynx or back wall of the throat, though not infrequently they begin in the nasal passages. The affection of the nose which is more apt to spread is nasal catarrh. The acute form of this disease will often times affect the larynx in a few days, while the chronic form will generally take a number of months. These phenomena are more often observed in singers afflicted with nasal catarrh, especially if the disease be localized in the large cavities of the back of the nose. I have so often seen the above result that I can almost guarantee it unless perchance the afflicted individual attend to the malady in time. The laryngeal affection thus produced is of a catarrhal nature and is especially marked in those wonderful little bodies the vocal cords and tissues in their immediate vicinity. The reason for the appearance of catarrh in the larynx as a secondary disease among singers is no doubt due to the constant use of the vocal bodies during vocalatory action. The causes for the development of throat diseases may be arranged under two headings, namely: Predisposing and exciting.

PREDISPOSING CAUSES.

By the term predisposing, we mean any cause, either in the system at large or in the immediate vicinity of the throat which renders the person more liable to the inception of these diseases.

The principal of the predisposing causes are four in number, namely:

OVERWORK, EITHER BODILY OR MENTALLY.

EXTREME SENSITIVENESS OF THE MUCOUS MEMBRANE LINING THE THROAT.

HERIDITARY CONSTITUTIONAL DISEASES, AS SCROFULA, SYPHILIS, ETC., AND

PREVIOUS THROAT TROUBLES WHICH HAVE NOT BEEN THOROUGHLY CURED.

When the constitution has been run down by overwork, the body is rendered peculiarly susceptible to the taking of disease.

In order that the system shall ward off the many affections which are constantly striving to prey upon it, it must be in a perfectly strong condition. When the system is weakened, the blood becomes more or less stagnant in the blood-vessels, which are found in great numbers on the surface of the body. The blocking up of the circulatory system, of course greatly interferes with its function, the distribution of the blood into every nook and corner of the body, and hence, in itself, is a cause of disease. With reference to the production of throat affections, the interference in the circulation of the blood, as above described, is manifested in the mucous membrane lining the several cavities located at the upper portion of the respiratory tract. Insufficient nourishment will also invariably place the body in a fitting condition for the reception of disease.

The vast army of sewing girls, who are alike overworked and underfed, furnish ample illustration of this cause.

Some individuals are endowed with peculiarly sensitive throats; it makes no difference how great care they take of themselves they are always "catching cold." In many instances this excessive sensibility is unavoidable but, on the contrary, in many other cases it is developed and maintained by the peculiar habits of the afflicted parties. Many individuals take altogether too great care of themselves; they accustom themselves to overheated apartments, seldom if ever venturing out of doors when the weather is at all inclement. If perchance these hot-house plants are compelled to venture out, they are so wrapped up as to be unrecognizable. What will be the result of such a course? Why, it will produce such an amount of delicacy and sensitiveness in the persons practicing it, that they cannot fail to take cold at the slightest exposure. It is astonishing what a trivial cause will produce a most serious cold in this class of individuals; simply standing by an open window for a few seconds, or sitting by a door slightly ajar, or a slight fall of temperature in the sleeping apartment during the night being oftentimes sufficient. Now, while I do not advocate unnecessary exposure, I should advise everybody to accustom themselves as far as practicable to the different climatic changes so characteristic of this portion of the globe. As regards clothing, the body should at all times be comfortably clad. The habit of bundling, more especially in the neighborhood of the throat, is a pernicious one and is, in itself, a prime cause of throat ailments. Frequently from neglect or bad treatment inflamatory affections of the throat which, had they been properly attended to at the time, would have been speedily cured, remain in a subdued or chronic form for an indefinite period. Although in many of these cases, the amount of inflammation remaining in the parts is very slight, yet it is sufficient to provide a nucleus for quite a severe throat trouble at the slightest cause. This well-known fact accounts for the many repeated attacks of sore throat with which some individuals are so constantly afflicted. I have seen and treated very many such cases, and I have repeatedly demonstrated the fact that when the throats of these persons are restored to their normal healthy condition, they are not more prone to these affections than their fellow beings.

Such heriditary affections as consumption, scrofula, syphilis, etc., render the afflicted party more liable to inflammatory diseases on account of the alteration in the constituents of the blood which these heirlooms produce. Some of the worst forms of throat affections which the specialist is called upon to treat are superinduced by the above causes.

EXCITING CAUSES.

By the term exciting we mean any cause whether climatic or otherwise which by its direct effect produces inflammatory action.

Among the most frequent of the exciting causes are:

THE DIRECT ACTION OF COLD.

THE INHALATION OF SOLID OR FLUID PARTICLES EXISTING IN THE ATMOSPHERE.

THE INHALATION OF TOBACCO SMOKE, AND THE PROMISCUOUS USE OF HOT AND COLD FOOD.

The manner in which exposure to cold produces inflammatory action is peculiarly interesting. The entire surface of the body is covered with an immense quantity of little glands which secret a fluid whose function is to keep the body in a moist condition. The fluid secreted in these minute glands which, as we all know, is perspiration, is poured out upon the body through an immense number of little tubes called the sweat pores. Now, if the body is subjected to an unusual amount of cold, all these little pores through which the perspiratory fluid oozes are instantly close, which in turn will cause an instantaneous checking of perspiration. When the cutaneous secretion is checked in the above manner, or any portion of the human frame, all the blood is driven from the surface to the organ or organs immediately subjacent, which causes them to be supplied with a greater amount of blood than is necessary, producing congestion the precursor of inflammation. If this checking of perspiration occurs in the chest, the trouble is manifested in the lungs — producing that most prevalent and fatal disease, pneumonia. If the sudden closure of the sweat-pores takes place in the lower portion of the back, kidney affections are apt to supervene, and finally, if the inter-ference to the flow of the perspiratory fluid takes place in the throat, laryngeal ailments are the result.

The reasons why the larynx and other organs in its immediate vicinity are so frequently attacked by cold are two-fold. *First*, their close proximity to the external surface of the body, and *second*, because they are located in that portion of the frame which is generally unclothed.

The first effect of the action of cold upon the mucous membrane lining the vocal organs is a checking of the natural secretion of the parts. This accounts for the dry and parched sensation which is expe-rienced in the mouth, and is so characteristic of the early stages of inflammatory affections in this cavity. The next effect is a more or less copious expectoration of a viscid and whitish-colored mucus. The latter symptom is manifested from twenty-four to forty-eight hours after the reception of the cold — that is when the inflammation has become more prominent. The inhalation of solid or fluid particles existing in the atmosphere under certain conditions, is also a most frequent exciting cause of throat difficulties.

These particles act either mechanically or chemically upon the structures with which they come in contact ; mechanically by producing irritation, which is the first step in the inflammatory process, and chem-ically by altering the nature of the parts they come in contact with.

Artizans exposed to the dust of various workshops, attendants in chemical laboratories and others similarly imperiled are most apt to suffer in this way. Inhalation of the dust of the streets, which as we all know contains many noxious and irritating ingredients particularly in windy weather, is also productive of many vocal ailments. The inhala-tion of tobacco smoke is also a prime cause of throat affections. It is not necessary for the sufferer to be a smoker himself, the mere fact of his being confined for hours in an atmosphere charged with the fumes of tobacco is frequently sufficient to produce these troubles. The above statement may seem a little too strong to the uninformed, but I have met with many cases of throat diseases which have been brought about in this manner during my long experience as a throat specialist. It is hardly necessary for me to add that the direct indulgence in the weed will oftentimes produce these ailments. In this instance they are caused by the constant irritation of the smoke, which in turn produces a dryness of the parts. The affections thus engendered are generally of a chronic or long-standing nature. The promiscuous use of hot and cold food and drink at the same meal is an apparent cause of throat difficulties. Thus we partake of hot soup, or drink hot coffee or tea, and cool the mouth by draughts of ice-water taken at intervals during the meal ; or after enjoying a warm dinner, we indulge in ice cream or water-ice, and follow this by a draught of hot coffee. This alternate application of hot and cold to the delicate mucous membrane of the throat can hardly fail if persisted in at least to place it in a condition favorable for the inflamatory process.

Prior to the invention of the laryngoscope the diseases peculiar to the vocal organs received little if any attention. Physicians in general were, apt to content themselves with prescribing some internal remedy, coupled with a gargle or some disagreeable external application. Some members of the medical profession, ignoring laryngeal affections altogether, were wont to ascribe their patients' symptoms to nervousness, billiousness, etc. Consequently the sufferer received very little sympathy and certainly no proper medication. Those, forsooth, who at that time undertook to treat the throat, did so in such a harsh and bungling manner that to make use of a common saying, " the cure was worse than the disease." A favorite method of treatment in the prelaryngoscopic period, was the introduction into the mouth of a sponge saturated with some strong caustic solution,'as for instance, nitrate of silver, or nitric acid. This plan of medication not only occasioned great pain to the patient, but in ninty-nine cases out of a hundred aggravated the disease.

Another favorite remedy in throat affections during the period above alluded to, was the depletion of the body by blood-letting, or venesection as it is technically called. It was thought at that time, by those who employed this latter plan of treatment, that if the amount of blood was lessened in the parts inflammatory action would be reduced. Another popular method of treatment then employed was salivation, that is the internal administration of mercury in such quantities as to produce inflammation of the mouth, especially in the neighborhood of the gums. In some cases salivation was carried to such an extent as to produce a loosening and a dropping out of the teeth.

Now, however, all things are changed, the treatment of diseases of the throat having been brought to such a state of perfection that it seems to me to be almost impossible to improve upon it. What has brought about this metamorphosis? Why, the laryngoscope, which by its wonderful powers have enabled us to study and understand the affections peculiar to the upper portion of the respiratory tract.

The treatment of the throat and neighboring parts, as employed at the present time consist mainly in the application of medicines directly to the diseased tissues. These applications are made through the agency of instruments especially designed for this purpose. No matter where the disease may be located whether in the remotest portion of the throat or nose it can be easily reached and the medicines thoroughly applied thereto.

Very many throat troubles which have resisted other methods of treatment for months, are oftentimes entirely cured by a few applications of drugs made in the proper manner.

The principal methods of treating the throat are three in number, namely :

THE MEDICATED SPRAY

BRUSH APPLICATIONS AND

INSUFFLATION

The medicated spray consists simply in the application of drugs in the form of a fine, spray into the diseased organs. The appliance made use of in the manufacture of these sprays is a peculiar one, and consists essentially of three instruments, an ordinary force pump, a copper boiler, and a set of glass tubes called spray-producers. Air is forced into the boiler by means of the force pump to which it is attached by an ordinary piece of rubber tubing. When sufficient air has been pumped into the boiler it is allowed to issue therefrom through a second piece of rubber tubing which is attached to the spray-producer. The spray-producer which resembles somewhat an ordinary cologne atomizer is inserted into a bottle containing the medicine to be used which, when the compressed air traverses the tube issues in the form of a very fine and powerful spray.

THE BRUSH TREATMENT consists simply in the introduction of small camels' hair brushes, saturated with some medicated solution into the diseased organ itself.

INSUFFLATION consists simply in the blowing of medicated powders into the affected parts. This operation is best performed through the agency of the apparatus above described in connection with the medicated spray. The tubes, however, although resembling somewhat the spray-producers are constructed on a different principle, since they are intended to transmit powders instead of liquids.

Of the several forms of treatment delineated above, the medicated spray is by far the most useful in the majority of throat ailments. For the treatment of that most prevalent of all diseases, nasal catarrh, this method is of the utmost worth, its employment being absolutely necessary to a cure of the disease.

In no other branch of medical sciences has quackery obtained such a firm footing as in the diseases of the throat and nasal organs. This state of affairs is not only due to the great prevalence of these affections in this climate, but also to the disposition of many individuals to exhaust every resource before applying to the skilled physician for relief. What is the result of this course? Why, very many troubles which could have been cured by a few applications and at a trivial expense if the patient had applied for treatment in time will oftentimes necessitate a long course of medication for their relief. This fact is especially true with reference to nasal catarrh for the cure of which disease, an innumerable number of instruments and infallible remedies have been from time to time invented and discovered.

These contrivances, if they are sufficiently well advertised are sold by the score and fill the pockets of the unscrupulous dealers. While many of the so-called cures are harmless in their action, there are many others positively injurious oftentimes destroying the delicate mucous membrane lining the upper portion of the respiratory tract.

I have many times seen extensive ulceration of the nose caused by the snuffing through the nostrils of powders guaranteed to cure this most distressing affection.

CHAPTER II.

NASAL CATARRH.

The diseases which are encountered within the throat and its immediate vicinity are exceedingly interesting, not only on account of their great frequency, but also from the evil effects produced by them upon the entire vocal apparatus. In the present discourse it will of course, be impossible for me to give a thorough exposition of the many ailments so frequently met with in this portion of the body. I have therefore endeavored to select those which on account of their prevalence among singers, shall be most interesting to this class of individuals.

The affections which I shall describe in the following pages are four in number, namely :

NASAL CATARRH.

PHARYNGITIS OR ORDINARY SORE THROAT.

LARYNGITIS and

PARALYSIS OF THE VOCAL CORDS.

The most prevalent by far of all the diseases found in the upper portion of the respiratory tract is nasal catarrh.

To no class of individuals does this affection occasion so much suffering and annoyance as to the singer who is in constant use of his vocal organs, since many vocal sounds are entirely dependent upon the healthy condition of the mucous membrane lining the nasal cavities.

Every cavity of the human body is lined with a soft and pliable membrane styled mucous membrane. It is called mucous because it secrets a fluid of more or less pasty consistency, the office of which is to keep the parts moist. I especially desire that this fact shall be well born in mind, because it applies equally well with reference to the mouth and the laryngeal cavity. The disease under consideration is simply a chronic or long standing inflammation of the nasal mucous membrane characterized by a more or less copious discharge of diseased mucous from the parts.

In well marked cases of this disease the membrane referred to above is often found to be the seat of large ulcers, brought about by the decay of the soft tissues of the nasal organ.

The symptoms of nasal catarrh are many and varied, the most prominent of which are :

OBSTRUCTION OF THE NASAL PASSAGES.

A COPIOUS DISCHARGE OF MUCOUS.

PAIN OF A DULL AND HEAVY CHARACTER AND IMPAIRMENT OF THE SENSE OF SMELL.

The obstruction of the nasal passages is due to a swelling or thickening of the membranous lining. In some severe cases of catarrh the nostrils are entirely occluded, consequently respiration through the nasal canals is completely shut off. This obstruction is usually greater in damp than in dry weather, and not infrequently we find that either one passage or the other is nearly wholly impervious to the air, there being no regularity with respect to the nostril affected. The constant discharge of fetid mucous is undoubtedly the most disagreeable feature of nasal catarrh since it compels the sufferer to make constant endeavors to discharge it. When the large cavity at the back of the nose, known as the post-nasal cavity is affected, there is a constant dropping of mucous into the back of the mouth. When the sufferer from nasal catarrh is unable to expel the mucous which is being constantly poured forth from the deceased membrane, the cavities of the nose, after a short time become completely blocked up. There are two reasons for this accumulation.

FIRST, The mucous being thick and tenacious becomes firmly attached, or, as it were, glued to the parts, and is therefore capable of resisting all ordinary efforts of expulsion.

SECOND, The mucous membrane being diseased is unable to perform its natural function, namely, the expulsion of the secretion accumulating on its surface.

Pain as a general thing is not a prominent sympton of this affection In some cases, however, it is well marked and the occasion of a great deal of worriment. Its most frequent seat is generally immediately beneath the eyes at the most depressed portion of the nose.

Deafness and impairments of the sense of smell oftentimes result from neglected cases of nasal catarrh. The first manifestation is due to the extension of the disease through the ear ducts into the cavity of the ear itself, while the second is owing to the deadening of the nasal mucous membrane.

Nasal catarrh in a vast number of cases is caused by frequent and repeated attacks of "cold in the head." Quite frequently the disease commences in a slow and insidious manner, and is present in the parts several months before the sufferer is conscious of its existence.

The influence of this disease upon the singing voice is at all times most marked, and may be either immediate or remote. The immediate effect of an attack of catarrhal inflammation will be the utter impossibility of rendering certain portions of the vocal register.

The tones most affected will be those which in their journey from the body issue solely from the nasal cavities. Those tones which issue partly from the nose and partly from the mouth will, of a necessity, be only partially affected by the above state of affairs. This interference to vocal sounds is due to the thickening of the mucous membrane lining the nasal canals.

In order that the thickening of the mucous membrane characterizing nasal catarrh be removed, it is necessary to undergo a series of treatments by the skilled physician which consists simply in the thorough and repeated applications of medicines to the diseased parts. The remote effect of nasal catarrh upon the voice is of the most pronounced character. When the vocalist thus affected begins to find out that certain tones are being poorly rendered, unless he thoroughly understands the cause, he is apt to strain every nerve to regain his lost power. By so doing he brings more force to bear upon those delicate bands, the vocal cords. For a time this method of singing improves the tones, but at a heavy price, namely, injury to the vocal cords. Why? Because the extra forces invoked are the stronger muscles outside the larynx — bodies which ordinarily have nothing to do with singing. What does this method of vocalization lead to? Strain of the voice which is greater or less according to the amount of nasal obstruction, and the length of time the extra forces are employed. This result I have seen very many times, in fact, scarcely a day passes but that a singer consults me who has been endeavoring to render tones which could not be made properly on account of obstructions existing in the nasal passages, the direct cause of catarrhal swelling.

Aside from its effect upon the vocal apparatus nasal catarrh is productive of much harm, and is a direct cause of many other throat complaints. Why? Because to a greater or less extent it compels the afflicted person to breathe through the mouth. The nose is the only channel through which air should pass into the body during ordinary acts of breathing, the mouth being intended only as an accessory breathing agent when on certain occasions—as for instance running—the lungs demand a rapid and increased supply of air.

The air in passing through the nostrils is warmed and sifted of its harmful ingredients and thus prepared for its reception into the delicate structures below. If it goes directly into the mouth without the above preparation, it will frequently cause irritation and inflammation of the mucous membrane lining of the mouth and throat by being, in the first place too cold, and in the second place, by containing irritating particles of dust and other matter.

We will now pass on to a consideration of that part of this subject which the singer, in all probability, will consider of the most importance, namely, can nasal catarrh be cured? My answer to this query is, certainly if the proper treatment be employed in the proper manner. The inquiring mind will immediately ask, why then do so many people, and among them many physicians, constantly assert that this affection is incurable? Simply because they have not properly investigated the subject. You must remember that the science of laryngology is still in its infancy, and that but a few years ago little or nothing was known concerning the cause and treatment of the many ailments peculiar to the throat, in fact, not until the discovery of the beautiful instruments by which we are enabled to see the hidden recesses of the throat and nose. Since the employment of this apparatus, diseases, which were hitherto not at all understood, have been investigated and their cause thoroughly made out. To none of these affections do the foregoing remarks apply more forcibly than to nasal catarrh, concerning the cure of which science has done so much during the past decade.

The treatment of this most distressing affection must be of the most thorough nature.

Every portion of the nasal mucous membrane must be treated with the proper applications and no part, no matter how remote, must be allowed to go untouched. The necessary applications are made through the agency of a special set of instruments which are manipulated under the brilliant light of the laryngoscope.

The most important and useful of the instruments used for this purpose is the compressed air apparatus. By means of this contrivance a most powerful liquid spray can be forced into the nasal cavities both in front and behind and the medicines thoroughly applied thereby. Of late years owing no doubt to the great prevalence of nasal catarrh, very many quack nostrums and patented appliances have been extensively advertised as specifics for the cure of this disagreeable affection. Each one of the above are always accompanied by an unlimited number of signed certificates which testify as to the inestimable value of the article in question and are thus calculated to attract the eye and the pocket of the unwary victim. I could, were I so disposed, fill a book with narratives concerning the effect of many of these articles upon the delicate mucous membrane lining the nose, but I shall content myself with the statement that the best of them are only palliative and not in any manner curative while the remainder are extremely injurious. Many of the nostrums contain ingredients the employment of which will not only increase the inflammatory action, but will also if persevered in, excite ulceration in the delicate nasal membranes. It stands to reason that no remedy, no matter how meritorious it may be, can effect a cure in a number of cases, since there are very many forms of nasal catarrh, and in order to differentiate between them it is necessary that a thorough examination be made by a skilled physician. Generally speaking that drug which may be most serviceable in one case may be extremely harmful in another. This assertion is not true only in the nasal organs but holds good with reference to ailments located in any other portion of the body. Some persons spend more money, and to no purpose, on quacks than the legitimate cure of their affection would cost in the hands of the regular physician. It is only necessary in this age of progress for the quack to prepare some nostrum, no matter how worthless it may be, and properly advertise it, and he is certain to reap a rich harvest gleaned from the pockets of the gullable public.

CHAPTER III.

PHARYNGITIS.

Pharnygitis or ordinary sore throat is simply an inflammation of the mucous membrane lining the pharynx or back wall of the cavity of the mouth. This disease may present itself in two forms, namely an acute and a chronic. In the ACUTE or recent form of inflammation the affected tissues are found, upon laryngeal examination, to be greatly swollen and congested the mucous membrane lining the parts presenting a deep red color. The sensations experience by the patient are peculiar, and characteristic of the disease. The principal symptoms are pain in swallowing, dryness of the mouth a harsh and rasping cough, and later on in the attack, a copious expectoration of frothy mucous. The most frequent cause of this form of throat trouble is exposure to cold when the body is unprepared to withstand it. Although very many cases of this disease are produced by the carelessness of the affected persons many more are caused by some unforeseen and unavoidable circumstance. Very many persons take cold in passing from a heated theatre or concert hall to the cooler street beyond. No matter how high the temperature may be outside it is generally from ten to twenty-five degrees warmer inside, hence the sudden transition will oftentimes produce quite a serious attack of sore throat unless, forsooth, the exposed party be provided with extra garments. In the chronic or long-standing form of sore throat the affected tissues are not nearly as red or congested as in the acute form, the symptoms, however, experienced by the patient are more numerous and more cleasly defined.

There is apt to be more or less difficulty in swallowing, the mucous membrane lining this portion of the throat being thickened and covered with a thick and pasty mucous. There is always present a sharp dry cough which is caused by the efforts of the patient to expel the mucous. There is also apt to be present a peculiar feeling as if the parts were covered with some foreign substance. The effect upon the voice of a chronic sore throat is sometimes quite marked. When the inflammation is great and there is much swelling of the tissues, there is a certain amount of huskiness and thickening of the voice with also a lack of vocal control and a tendency to tire easily. Among the causes for this disagreeable affection, there are two which deserve special consideration, namely :

THE IMMODERATE USE OF TOBACCO, AND THE ABUSE OF ALCOHOLIC LIQUORS.

It is an undisputed fact that the smoke arising from tobacco is an irritant to the delicate mucous membrane lining the air passages. This effect is more especially noticeable in persons afflicted with delicate throats. It has been asserted as an argument in favor of the use of tobacco, "That the mucous membrane of the throat becomes, after the lapse of a certain period of time, hardened and inured." This is merely a possible result and should not induce the singer, especially if he have any throat trouble, to persevere in the use of tobacco. The vocalist who wishes to preserve his voice should not smoke, but if he must use the weed, let him remember that smoking immediately after singing is harmful, because the vocal organs are then in a congested state and easily acted upon by any irritant.

There is another fact with relation to this subject and which should always be born in mind by those indulging in the excessive use of tobacco, and that is that the constant absorption of nicotine, which, as we all know, is deadly poison, is not only injurious to the system at large, but also to the delicate tissues of the mouth. I cannot too forcibly impress it upon the mind of the singer, if he wishes to retain his voice in its natural purity, to abstain entirely from the use of tobacco, even though for the time being it may seemingly produce no ill effect.

If the employment of tobacco is harmful to the vocal organs the use of alcoholic stimulants is ten-fold more so.

Alcohol is an irritant to all mucous membranes, especially to that lining the air passages, and if used to any extent will assuredly in time create inflammatory action in these parts. If the drinking habits are persevered in sufficiently long permanent injury will undoubtedly be done to the vocal apparatus.

Some artists, prompted no doubt by bad advice, are in the habit of imbibing alcoholic stimulants immediately previous to vocalizing. I wish particularly to warn singers against this pernicious habit which, if persisted in, cannot fail to be productive of serious consequences.

The treatment of the two forms of sore throat depicted above must be chiefly of a local nature : that is to say the medicines employed must be applied directly upon the diseased surfaces. The most valuable agent in the treatment of these ailments is the compressed air apparatus already incidentally alluded to. By means of this appliance the medicated solutions can be thoroughly distributed over the diseased tissues in the form of a fine and powerful spray.

Camel's hair brushes attached to a long and slender handle can also be often used to advantage in making applications in this species of throat trouble, but their employment, except in some special cases, is neither as agreeable nor as efficacious as the compressed air spray. The other forms of treatment generally employed in this class of diseases are gargles and medicated troches. These agents are in no manner curative, they are simply pallative. They are beneficial in two ways, namely : By keeping the parts free from mucous and by soothing the inflamed tissues.

CHAPTER IV.

LARYNGITIS.

Laryngitis is an inflammation of the mucous membrane lining the larynx or voice-box. It presents itself like the affection just considered, in two forms, namely: An acute and a chronic. The acute or recent form is generally superinduced by exposure to cold or rapid changes of temperature. Upon inspection of the parts with the valuable assistance of the laryngoscope, the larynx presents a characteristic appearance, being very red and swollen, a direct effect of the inflammatory process. These manifestations are generally most noticeable in the vocal cords the sound-producing reeds of the voice-box, these bands, being very highly colored, thus offering quite a contrast to their natural appearance, namely: A pearly white. The most prominent symptom, however, is impairment of the voice both in talking and singing. The talking voice is husky or hoarse in direct proportion to the severity of the attack. When the voice is thus affected talking in a loud tone, even for a short time, will invariable increase the hoarseness. The singing voice is always greatly affected; proper vocalization being out of the question. The reason why the voice is affected by this form of laryngitis is easily demonstrated. The vocal cords, in order to perform their function properly, must be in a healthy condition, that is to say they must possess their natural proportions. When they are the seat of inflammatory action they become thickened and congested, which not only destroys their vibratory power, but also to a greater or less extent obliterates the chink of the glottis, the opening between the two cords which is for the purpose of allowing the air to escape through the laryngeal tube. This closure of the chink of the glottis is a prominent cause of vocal impairment because the air, which passes through the voice-box in a healthy condition, is the motor power which causes the vocal bodies to vibrate. Cough is also a prominent symptom and is caused in two ways: First, by a constant tickling of the throat, and second, by the copious secretion of a thin mucous. The cause of acute laryngitis is, like most of the other inflammatory affections of the throat, exposure to cold and dampness.

It is sometimes surprising what a slight exposure will produce this affection, the afflicted one being oftentimes totally unaware of having exposed himself. The phenomena of taking cold is decidedly interesting and should in a measure be understood by everybody, especially those constantly using their voices.

When the body is overheated, its surface is covered throughout its entirety with profuse perspiration. Now if while in this condition it be subjected to cold, all the minute sweat pores which abound on its surface and through which the perspiratory fluid oozes, will be instantly closed, which in turn will cause an instantaneous checking of perspiration. When the cutaneous secretion is checked in the above manner on any portion of the frame, all the blood is driven from the surface to the organ or organs immediately subjacent, which causes them to be supplied with a much greater amount of sanguinious fluid than is necessary, producing congestion, the precursor of inflammation. If the system is in prime order it will generally be able to combat successfully this congestion and prevent its development into inflammation, but if it is weak and run down, inflammatory action is a foregone conclusion. The reasons why the larynx, and the other organs in its immediate vicinity, are so frequently attacked by cold are two-fold: *First*, they are very near the external surface of the body, being covered in lean persons by the skin and a few ribbon-like muscles; *second*, they are situated in that part of the frame which is, as a rule; unclothed.

There is a time when the singer is especially liable to take cold, and that is when he proceeds from a warm apartment into the colder atmosphere beyond immediately after acts of vocalization. The larynx when at work requires a greater amount of blood than during rest, this causes it for the time being to be in a congested state, which congestion, however, is perfectly natural. When the larynx has accomplished its task and is quiescent, the above natural congestion gradually subsides, until the vocal organs contain only their normal quantity of blood. Now, if the vocalist should expose himself to the cold street air before this congestion has entirely subsided, he is almost sure to suffer from his indiscretionary act. The manner of dressing has a great deal to do with the health of the vocalist. The body should be clothed sufficiently for warmth and comfort. Too much clothing is as bad as too little because it produces an overheating of the body which in time causes a free perspiration, a state of affairs, as we have seen, extremely favorable for the reception of a cold. The overbundling of any portion of the body, particularly the chest and neck, with wraps, mufflers, &c., renders the wearer peculiarly liable to colds because it produces an extra sensitiveness of the parts. The minute a person thinks he has taken cold he piles on chest protectors, extra wraps, &c., not forgetting to encircle his neck many times with red flannel. This procedure can do no possible good but the greatest amount of harm for the several reasons cited above. A very prevalent way of taking cold is the tarrying in overheated apartments with heavy clothing on. This habit is especially noticeable among the ladies during their numerous shopping excursions. They will remain sometimes an hour or more in a store heated to a temperature of from 80° to 100° without removing their sacks and then, while their bodies are bathed in perspiration, proceed immediately to the cold street beyond. These remarks are especially applicable to those wearing seal skin sacks, which have a special tendency to the overheating of the body.

Singers, more especially those having delicate throats, should take care to keep the feet warm and dry. Some people are so sensitive that a wetting of the feet, no matter how slight, will invariably be followed by a cold.

CHRONIC LARYNGITIS.

This form of laryngitis is an inflammation of the mucous membrane lining the larynx, of a mild type and chronic character. This, of all the laryngeal affections, is the most prevalent amongst singers. The most prominent symptoms are a feeling of tickling or irritation referable to the larynx and impairments of the voice either of the singing, or of the talking, or of both. There is generally an irritating cough with a constant desire to clear the throat. Upon inspecting the parts with the laryngoscope, the entire mucous membrane lining the larynx will be seen to be congested and considerably thickened in certain localities. The vocal cords themselves will be found much thicker than normal and reddened instead of being pearly white as in health. The characteristic redness is apt to manifest itself in spots or streaks on the vocal bodies. A quantity of frothy mucous is nearly always present on the diseased tissues. In singers and professional speakers chronic laryngitis is apt to be localized in certain portions of the larynx the vocal cords being the chief point of attack. There is one condition quite often seen in the throats of vocalists affected with this disease, and that is a relaxation or giving way of the cords from over-exertion or strain of these bands.

Chronic laryngitis frequently occurs, as a result of a cold, or more properly speaking, a series of colds. Among singers, however, this affection has several special causes, the principal of which are :

1. Improper training.
2. Injudicious singing.
3. False singing.
4. Strain of the voice.
4. Other throat affections.

IMPROPER TRAINING.

We cannot over-estimate the evil effects produced on the delicate structures of the larynx by bad training of the voice. I have frequently seen a larynx which was perfectly well prior to teaching become utterly destroyed in a course of vocal lessons extending over a period of from two to three months. One of the chief ways in which chronic inflammation of the larynx is produced in singers, is the practice in vogue with a certain class of teachers, of striving by rapid methods to develop the voice in a short time. The development of the voice, as every good instructor fully knows, is a tedious affair and can only be accomplished after years of study. I know that rapid advancement is the tendency of the age, and that by developing a voice in a short time the teacher will cater not only to the pupil but also to his friends; but still the fact must never be lost sight of, that by endeavoring to develop the vocal apparatus rapidly, permanent injury is apt to follow. How does over-training produce chronic larynx? By making use of the stronger laryngeal muscles, bodies which ordinarily have nothing to do with vocalization. Why is the aid of these external laryngeal muscles invoked? Because the delicate vocal muscles not having had time to develop sufficiently to perform the duties imposed upon them, outside help is sought. The direct effect of the use of these stronger muscular bodies manifests itself in the vocal cords. In the first place, these bodies are congested, that is to say a large amount of blood is drawn to the parts by the extra force used. The next step in the process is the production of swelling in the parts especially in the neighborhood of the cords and finally after the lapse of a varying period of time, which depends upon the exertions of the teacher and the susceptibilities of the pupil, this congestion and swelling becomes a chronic inflammation and thickening. The mistaking the register of the pupil will oftentimes produce a chronic laryngitis in a manner somewhat similar to that described above. Unskilled teachers will frequently attempt to develop a tenor voice from one which nature intended to be a baritone or produce a soprano where there are only the qualities of a mezzo-soprano. This is also often done by unscrupulous masters in their endeavors to please a certain class of people who judge a voice by the height to which its possessor can go. It is in the rendition of the higher notes of the register that the efforts of the singer will produce the species of inflammation under consideration. Why? Because in the rendition of these notes a greater muscular action will have to be brought to bear on the vocal cords because the latter bands are not of the proper conformity to produce the desired tones with the unaided efforts of the delicate vocal muscles. Generally speaking the higher the voice the thinner or finer will be the vocal cords. This state of affairs is to allow of a greater velocity in the vibration of these little bands, for the greater the number of vibrations in a specified time the higher will be the resulting tone. Now when the cords are thicker as in a mezzo-soprano in the female, or a baritone in the male, in order to be made thin enough to produce tones that properly belong to the soprano or tenor, the cords will have to be stretched more than the delicate vocal muscles are able to do, hence the stronger or outside laryngeal muscles are made use of, the action of which bodies produce the same diseased conditions enumerated with reference to over-training. The following case clipped from my note-book will serve as a good illustration of this subject. Mr. G——, a young gentleman aged about twenty, consulted me in January, '82, to ascertain if possible the reason of his inability to sing. Upon laryngeal examination both vocal cords were seen to be congested and thickened, and the tissues in the immediate vicinity considerably inflamed. There was likewise visible in the larynx a large amount of mucous, which, as I afterwards ascertained, occasioned a great deal of annoyance from cough. These symptoms, as we have already seen, are those of chronic laryngitis and such was my diagnosis. Upon making inquiry, with a view of ascertaing in the cause of the trouble I elicted the following facts. About six months prior to his consulting me, my patient had placed himself under the tuition of a singing master of considerable repute in this city.

Upon a cursory examination of his pupil's voice, the teacher set him down as a tenor and forthwith proceeded to train him as such. After the taking of a few lessons, the pupil noticing that it was exceedingly difficult for him to render certain tones, so informed his teacher, who in response told him that it was of no consequence, but that if he persevered all would turn out right in the end. My patient being of a persevering and enthusiastic nature, kept on in his endeavors, his master all the time compelling him to render tones which he was totally incapable of making without forcing the voice. Matters kept on in this way for a time, the pupil gradually losing his vocal powers, until fortunately the voice broke down altogether rendering further tuition impossible. Such was the condition of affairs when the sufferer applied to me for advice. After listening to the above narrative and informing my patient as to the cause of his trouble, he placed himself in my hands for treatment.

His disense yielding readily to treatment, I discharged him cured, after the lapse of several weeks, with the strictest injunction not to use his vocal organs for at least six months. I did not see the gentleman again until the following December, when he called to have his vocal apparatus examined to see if everything was all right. Upon a thorough examination I found the parts perfectly healthy in every particular, the vocal cords themselves being much stronger than when I last saw him. He then informed me that he had begun tuition again in September and that his teacher (not the former one) recognizing his voice as that of a baritone, was training it as such with very good results.

INJUDICIOUS SINGING.

Under this heading a great variety of causes may be arranged as producing chronic laryngitis, I shall, however, confine my remarks to the most interesting one of all, namely : the exercising of the voice when it is husky or hoarse. I think, nay, I am quite certain, that nine-tenths of the cases of chronic laryngitis among vocalists are produced

in this manner. When the voice is husky or hoarse it is always a sign
that the vocal cords are more or less congested. Now, as I have
already stated elsewhere, when these bodies are to any degree inflamed,
in order that such inflammation shall subside, it is necessary that the
parts be kept quiet. As we all know in order to keep the cords still it is
necessary to refrain from acts of vocalization, since vocal sounds are
chiefly produced by the direct action of the vocal cords. Sometimes
the vocalist from a misunderstanding of the cause of the difficulty, and
sometimes from bad advice, exercises his voice. No efforts of vocalization, no matter how guarded they may be, will ever have a tendency to
remove laryngeal inflammation, but, on the contrary, will always increase
it, and if persevered in, will eventually produce a chronic inflammation
of the parts. Quite frequently vocalists are compelled when under an
engagement to vocalize during the existence of hoarseness and, sooner
than compromise their positions, they use their voices when they know
it will hurt them. One such indiscretionary act may not harm the voice,
but a persistence in such a course will invariably destroy it. If singers
will stop for an instant and think of the danger they run in pursuing
the above plan, they will be more reconciled to a little temporary
loss of vocal power, and more willing to give their throats the needed
rest.

FALSE SINGING is also another prime cause of chronic laryngitis. The term false singing is a comprehensive one; its general
significance, however, is the employment of false or improper vocal
methods. The undue employment of the falsetto register may also be
arranged under this heading. Of the many and bad methods employed
by poor vocal teachers, that which instructs the pupil to keep his larynx
fixed during the producting of the many and varied tones so characteristic of vocalization, is perhaps the most pernicious. The advocates of
this peculiar "method" claim that by so doing, not only is the power of
the voice enhanced, but that its brilliancy is increased. Whether this
is so or not I am unable to state ; but this I can assuredly state, that
this "method," so-called, if persevered in for any length of time will. if
it does not permanently injure the vocal cords, engender faulty habits of
singing. which will take much time and labor on the part of a good
teacher to correct.

In order to hold the larynx in one position during vocalization, the
large muscles attached to the outside of the organ are brought into
action, and these bodies acting directly against the vocal muscles,
impose a much greater force on the vocal bodies than should under any
circumstances be employed. It is this extra force which does the harm
and produces the result tabulated above. In the production of the
falsetto register the vocal cords do not vibrate throughout their entirety,
as in the singing of a full chest tone, but only in their free edges.

The cords are, as it were, held in a sort of vice which allows the air current, the moter power, to exert an influence only on the edges of the vocal bodies. This action on the part of the vocal bands is, to my mind, improper and will in time produce a thickening and congestion in the voice-box.

STRAIN OF THE VOICE.

Strain of the voice is a most frequent cause of chronic laryngitis. By straining the voice we mean making it perform work that it is utterly incapable of doing properly.

Whether the harm is done under the guidance of a teacher or whether by the unaided efforts of the pupil, the result is just the same, namely : A diseased condition of the larynx. The delicate little bands, the vocal cords, will not stand any nonsense, hence they rebel when treated improperly, a species of congestion being excited which produces hoarseness. Now, when this hoarseness is produced, if the singer should recognize the cause and remedy it, very little harm would be done; but, I am sorry to say, the reverse is generally the case, the straining is persevered in until, what was at first an ordinary congestion and swelling, has resolved itself into a chronic inflammation and thickening.

There is a golden rule which, if always observed by singers, would preclude the possibility of injury being done to the vocal cords, namely : " That all acts of vocalization which are productive of or followed by hoarseness, be it ever so slight, are improper and certain to produce a chronic inflammation or other injury, if persevered in."

OTHER THROAT EFFECTIONS.

Diseases located in other portions of the vocal apparatus are often indirectly the cause of chronic laryngitis.

An ELONGATED UVULA will frequently do this. This little organ when diseased is oftentimes so long that it hangs into the cavity of the larynx itself and keeps the tissues of this organ in a constant state of irritation. This irritation will, after a time, occasion congestion which in turn will produce the disease under consideration. If the laryngitis thus excited has existed for a short time, the simple operation of exercising the Uvula will relieve it; if, however, the inflammation is deeper seated, this operation will have to be followed by a course of medical treatment.

ENLARGED TONSILS will also in a measure be productive of chronic laryngitis in the singing voice. These glands when enlarged offer quite an obstacle to the proper action of the voice. In order that any sort of a tone shall be produced, in persons suffering from enlargement of the tonsils, a greater force must be exerted on the vocal bodies, which as we have repeatedly seen is productive of laryngeal inflamation.

NASAL CATARRH, especially when the obstructions are marked in the nasal passages is conductive to the above form of laryngeal trouble in almost the same manner as enlarged tonsils, that is by offering an obstacle to the escape of certain tones, those traversing the nasal canals.

What will eventually be the result if chronic laryngitis is allowed to remain? Why an entire destruction and breaking up of the voice. The time for this final giving out of the voice, will depend entirely upon the manner in which the vocal apparatus is used. If it is treated badly, a few weeks will be sufficient, but if, on the contrary, the voice is used carefully it may last for several months.

The treatment of laryngitis as with the other forms of throat disease previously described, must be entirely of a local nature. The medicines used must be applied into the larynx itself through the agency of instruments especially designed for that purpose. There are three methods of making applications to the interior of the voice-box, namely:

The compressed air spray method.

The brush method, and finaly

Insufflation.

The spray treatment is essentially the same as that described when speaking of pharyngitis, with the single exception that the spray tube points downwards instead of forwards. The brush method consists simply in the introduction of camel's hair brushes attached to a long and curved handle into the interior of the laryngeal organ. Insufflation consists simply in the blowing into the larynx of medicated powders by an instrument which is called an insufflater.

Many singers have an idea that the direct application of medicines to the vocal apparatus is productive of great pain. This is a great mistake, for when the operation is performed by the skilled manipulator very little annoyance is occasioned and in some cases, particularly those of nasal catarrh, the treatment is rather agreeable than otherwise. Of course if a person who does not understand the throat attempts to treat it, much pain will result from his efforts *first*, because of a lack of skill in handling the instrument, and *second*, on account of the medicines employed, these persons generally using much too powerful remedies.

CHAPTER V.

PARALYSIS OF THE VOCAL CORDS.

Paralysis of the vocal cords is by far the most interesting of all the types of laryngeal disease. In order to fully understand the description of the several paralytic affections to be shortly given, it will be abso-

lutely necessary to stop for a moment and look into the physiological
action of the vocal cords. As I have fully treated of this subject in my
work entitled " The Throat in Relation to singing," I shall merely give
a brief synopsis here. Vocal sounds are produced by the vibration of
the vocal cords, which action is caused by the passage of the air-current
through the chink of the glottis, the opening between the two vocal
cords. What does this vibratory action on the part of the vocal bodies
do? Why it divides the current of air into a large number of little
currents which are then known as tone-waves. These tone-waves
passing up into the mouth are formed into articulate speech by the
action of the tongue and other organs contained in the mouth. The
vocal cords are merely two thin elastic bands of tissue running across
the larynx from before backwards, attached firmly in front to the inside
of the voice-box immediately behind the prominence known as the
" apple," and behind to two bones which are called the arytenoid carti-
liges. The vocal cords together with their points of attachment to the
larynx are beautifully displayed in the subjoined cut.

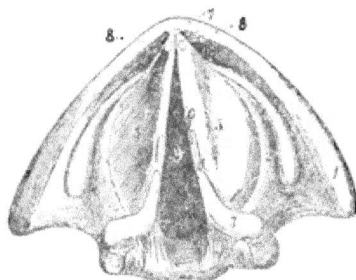

Figure 1. Horizontal section of the Larynx.

1. Outer framework of larynx. 3. Arytenoid cartilages, the bones
to which the vocal cords are attached behind. 5. Vocal cords. 7. The
point of attachment of vocal cords in front.

In this drawing the cords are represented as being open as during
inhalation; when they are brought together the passage between them
which is then called the chink of the glottis, may be represented by a
single dark line. The bones to which the cords are attached behind
the arytenoid cartilages, are of the greatest importance since it is mainly
by their action that the cords perform those movements, so characteristic of them and upon which vocalization principally depends. Since the
principally movements of the vocal bodies are dependent upon the
action of the arytenoids it follows of a necessity, the muscles which
preside over the vocal movements, must be attached to these cartilages.
This indeed is true with a single exception which shall be presently
alluded to. When the cords are separated as during respiration and it
is desired to use the voice these bodies are brought together. This

action is known as adduction and is produced as you may easily imagine by simply drawing together the two arytenoid cartilages. This action is performed by a muscle which runs from one arytenoid to the other, and is attached to both bones. This body, which is displayed in figure 2, is called, from the effect which its action has on the vocal cords the adductor muscle, a derivation from two Latin words, *ad* and *ducto*, meaning to lead to.

1. Epiglottis, or guardsman of larynx; 4. cricoid cartilage, the body on which the arytenoids revolve; 5 arytenoid cartilages. T. transverse laryngeal muscle the adductor of the vocal cords P. abductor muscle or separator of vocal cords.

Whenever it is necessary to take a breath during vocalization the cords must fly back or separate. This action is called abduction which term is a derivation from two Latin words *ab* and *ducto* meaning to lead from. This action is brought about through the agency of a single muscle, fig. 2. P.) which is attached as plainly shown in cut to the lower part of the ary- tenoid and the back of the cricoid cartilage. This muscle separates or

Fig. 2.

Framework of the larynx seen from behind.

abducts the cords by simply revolving the arytenoids outwards. In the above drawing the left muscle is only depicted, their being a similar muscular body on the opposite side. There are two other actions of the vocal cords, making four in all, which remain to be described, namely: tension and relaxation.

TENSION is performed by a single muscular body called the ten sor muscle. It is attached to the front of the voice box on the outside and tenses the cords by drawing the front of the larynx directly down- wards which increases the distance between the two points of attach- ment of the vocal cords in front and behind and stretches these banes

RELAXATION is performed by two muscles style the relaxors and which are attached to the same bones as the vocal cords. There is one of these muscular bodies for each vocal cord and they relax these bands by drawing together the two portions of the voice-box to which they are attached.

PARALYTIC AFFECTIONS of the vocal cords may be most conveniently divided into four varieties, each one of which shall corres-

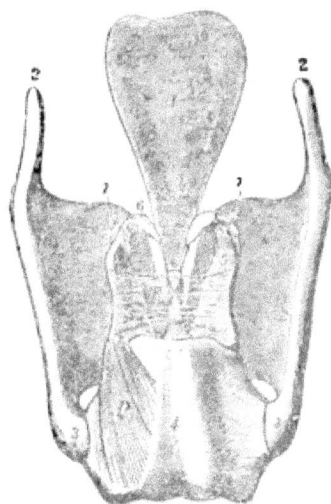

pond to one of the four movements described above as characteristic of the vocal bodies.

These forms enumerated in their respective order are :

1. Paralysis of adduction.
2. Paralysis of abduction.
3. Paralysis of Tension.
4. Paralysis of Relaxation.

PARALYSIS OF ADDUCTION is a lack of power in the adductor muscles to draw together the arytenoids and thus adduct the cords. In this variety of paralysis which is shown below the vocal cords are permanently separated from inability on the part of the adductor muscles to approximate them.

a. a, the two arytenoids, h, back wall of larynx s, s, points of attachment of cords to arytenoids, c, opening between the cords, k, attachment of two cords to front of larynx.

The annexed illustration, which is simply an outline drawing, will give a pretty good idea of this variety of laryngeal paralysis. When an attempt at vocalization is made the vocal cords instead of approximating or coming together, as they naturally would in a healthy larynx remain permanently fixed at the sides of the voice-box. Why ? because the transverse muscle (see fig. 2) which presides over abduction, being paralyzed, is unable to perform its duty, namely, the drawing together of the arytenoids and with them the vocal cords.

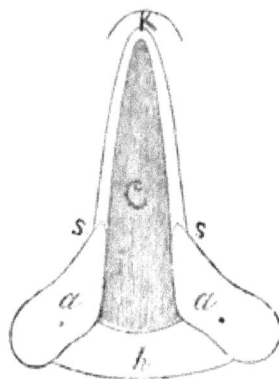

Fig. 3.

Paralysis of adduction of vocal cords.

What will be the effect of this form of laryngeal disease upon the voice ? A complete loss thereof, both in speaking and singing ; because the cords not being able to come together there can be no vibration in them. This, which is the most prominent symptom of paralysis of adduction, is called aphonia. The only other symptoms liable to manifest themselves in this affection are a slight cough and irritation due to a congestion of the parts.

PARALYSIS OF ABDUCTION is simply an inability on the part of the abductor muscles or separators of the vocal cords, to abduct or separate these bodies.

In this variety of paralysis the two cords remain permanently fixed in the middle of the larynix as shown in Fig. 4.

A. A. arytenoid cartilages ; B. B. paralyzed cords ; C. front wall of larynx ; D. back wall of larynx.

The cords occupy the above position because the adductor muscles, acting properly, bring together the cords, but the abductor muscles being inactive on account of paralysis are unable to bring the vocal bodies back to the sides of the larynx. The two sets of muscles, the abductors and the adductors, when healthy are antagonistic to each other, consequently when either set is paralyzed, there being no opposing forces brought to bear against the healthy set it acts continuously, thus when the abductors are paralyzed, the adductors keep the vocal bodies permanently united.

The muscle involved in paralysis of abduction is depicted in fig. 2. (P). It, as we have seen, abducts the cords by revolving outward the arytenoid cartilages. In order that air shall pass through the larynx in its journey to the lungs, it is absolutely necessary that the vocal cords be widely apart. If this condition be not permitted and the cords are drawn close together, no matter how slight may be the degree of such approximation, the proper performance of respiration is interfered with. The most prominent symptom of this form of laryngeal paralysis is a great difficulty in breathing or dyspnoea as it is technically called. When the paralysis is complete, that is to say, when the cords are in direct opposition, the above interference to respiration is marked, and the occasion of the greatest suffering, all the air that enters the lungs having to pass through the narrow opening known as the chink of the glottis. The voice is not apt to be much interfered with, because the vibratory power of the cords is not destroyed, and because these bodies are approximated in the proper manner to produce tone. The difficulty of breathing above described comes on in spasms, as it were, during each of which the life of the sufferer is seemingly in great danger. After a while inflammatory action is set up in the larynx which, on account of the accompanying swelling, greatly increases the sufferings of the patient. The above symptoms increase until finally, unless relief is afforded by surgical interference, the sufferer must die. Fortunately paralysis of abduction, as depicted above is an extremely rare affection, this disease being generally of a partial nature, and its symptoms therefore somewhat modified.

Upon inspecting the larynx in a severe case of paralysis of abduction, the picture presented upon the laryngoscopic mirror will be most striking. The vocal chords will be seen drawn together and lying almost

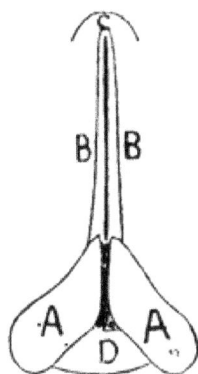

Fig. 4.

Paralysis of abduction
of vocal cords.

motionless in the centre of the larynx, and the opening between these bodies, the chink of the glottis will be clearly defined and about a line in extent. The larynx will also be seen to be the seat of inflammatory action, the mucous membrane being quite red and somewhat swollen.

PARALYSIS OF TENSION is an inability on the part of the tensor muscles to stretch the vocal cords. In this variety of laryngeal paralysis the vocal cords, when adducted present a relaxed condition from a lack of action in the muscular body presiding over tension. Adduction and abduction are in no wise interfered with, the vocal cords being approximated and separated in a perfectly normal manner. Upon laryngeal inspection the picture presented in this affection is somewhat like the subjoined cut.

A A, arytenoids; B B, relaxed cords; C, front wall of the larynx; D, back wall of larnyx; E, ellipitical opening between cords.

In order to thoroughly understand this form of laryngeal paralysis, as well as the variety next to be described, it will be necessary to glance for a moment at the effect exerted upon the singing voice by the two vocal movements called tension and relaxation. When the two cords are brought together by the phonatory act, a tone located about the middle of the vocal register can be rendered without much action on the tensor or relaxor muscles. If, however, a higher note be demanded, in order that the vocal bodies shall make the required number of vibrations they must be made tensor. This duty is performed by the tensor muscles,

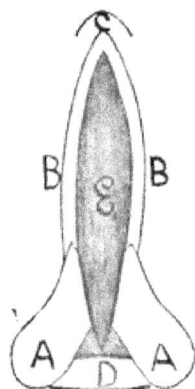

Fig. 5.
Paralysis of tension.

which bodies stretch the cords as we have already seen, by pulling the front part of the larynx downwards. The higher the note is placed in the scale the more will it be necessary to tense the cords.

Now, supposing when the cords are brought together a lower instead of a higher note is required what then must transpire? Why the cords must be relaxed, for the reason that the number of vibrations necessary to produce a lower tone is less, and that in order to diminish the number of vibrations in any giving period of time, the aid of the relaxor muscles must be invoked. The lower the note the slower must be the vibrations, and consequently the greater must be the relaxation of the vocal bodies. When, therefore, the tensor muscle is paralyzed and unable to perform its function, that is the tightening of the vocal cords it will be impossible for the affected person to render, in any degree, the higher notes of the register.

PARALYSIS OF RELAXATION is an inability on the part of the relaxor muscles to perform their function, namely the relaxation or loosening of the vocal cords. In this species of vocal paralysis, while the cords can be brought together and separated in a perfectly normal manner it will be impossible to relax them. The relaxor muscles, which are called the vocal muscles par-excellence, relax the cords by drawing together the two points of attachment of the vocal cords, which action of course loosens these bodies. In the above movement, the relaxor muscles are directly antagonistic to the tensor muscles. Now supposing from paralysis or other cause, the relaxor muscles are unable to relax or loosen the cords what will happen? Why the tensors having no opposing forces to contend against will keep the cords in a perpetual state of tension.

As you have undoubtedly already suspected the effect upon the voice by the variety of laryngeal paralysis under consideration will be an inability to render the lower notes of the register; because the vocal cords cannot be relaxed enough to allow of a sufficiently slow vibration for the production of these tones. I had, several years ago, a remarkable and instructive case of this form of laryngeal paralysis which I think will be interesting to the singer.

Mr. H., aged about 35, consulted me October 16, 1879, at the suggestion of his vocal teacher, to obtain an opinion regarding a peculiar defect in his voice. Upon interrogation I learned the following history. About two years previous, after an attack of cold, Mr. H. noticed that in talking the tone of his voice had altogether changed; that while previous to his " taking cold " he could talk in the ordinary tones of an adult male, he was subsequently unable to lower his voice, but was compelled to speak in his upper register; in other words, his voice was the *fac-simile* of a boy's prior to the change which occurs at about the age of fourteen. The singing voice was likewise similarly affected, it being impossible for him to render any of the notes above the middle register. Thinking that time would mitigate the difficulty, my patient pursued no course of treatment other than an attempt to regain his lost powers by vocalatory exercises. This latter procedure being entirely void of good result, he was induced to seek medical advice. At the time of seeing Mr H., the difficulty above described was so manifest that were I to have conversed with him a dark room I would most certainly have pronounced him a boy of about twelve or thirteen. Upon laryngeal examination I found the larynx comparatively healthy, there being but a slight amount of congestion in the tissues adjacent to the cords. Upon requesting the patient to phonate I noticed that although adduction was performed perfectly, the vocal chords were both in a state of great tension—such a condition in fact as one would expect to find during the rendition of the higher notes of the register. Again and again I repeated the experiment, requesting the patient to render certain musical tones, with a similar result, each successive examination confirming its predecessor. The most curious circumstance connected with the whole case was the following : by pressing on the front of the larynx in the neighborhood of Adam's apple, the voice suddenly assumed a natural character, dropping from the tones of a boy to those of an adult male. As long as this pressure was kept up Mr. H. talked quite easily in what would naturally have been his voice, were not his larynx the seat of the paralysis of the relaxor muscles, but the moment the hand was removed from the throat, the voice assumed its former unnatural condition. The reason of the dropping of the voice when pressure was exerted on the voice-box, was that this action took the place of the paralyzed muscles and relaxed the cords by diminishing the distance between the points of attachment of the vocal bodies. The causes of the several forms of paralysis enumerated above are oftentimes involved in obscurity. Mental emotion will sometimes produce them ; instances of sudden loss of voice through fright being quite common. The variety of laryngeal paralysis described under the head of paralysis of abduction, and which is characterized by great interference with respiration, is generally caused by pressure on the laryngeal nerve which thus interferes with or shuts off the nervous supply to the larynx. Overwork is a frequent cause of laryngeal paralysis in the vocalist. Many great singers have lost their voices at some time or another by this practice. The immense price sometimes paid to these artists stimulates them to sing through a long season and many times when they are utterly unfit to perform. I know of several artists now very popular, who are doing too much vocal work, and which will end up by an entire destruction of their voices unless they take the much-needed rest. Since the reign of comic opera many have risen and fallen, and this will continue to be the case as long as the singer is compelled to vocalize six nights a week and two matinees. No artist who has a good voice should endanger it by such continuous action ; it is too much to expect of the delicate vocal cords, and it will assuredly tell upon them in the long run ; it is simply a matter of time.

STRAIN OF THE VOICE is another frequent cause of vocal paralysis. This is often the result of vocalizing with the cords at too high a tension, or in other words attempting to sing entirely outside of the normal register.

Some forms of vocal paralysis, especially that type described under the heading of paralysis of abduction, are exceedingly difficult to cure. Those varieties, however, which are so often found in the throats of singers, as a direct result of abuse of the voice, are amenable to treatment and oftentimes speedily cured.

As in paralytic affections of other parts of the body, electricity forms the sheet anchor of hope in the treating of laryngeal paralysis.

The method of its administration is most unique, and simply consists in the introduction into the larynx itself of what is known as a larngeal electrode.

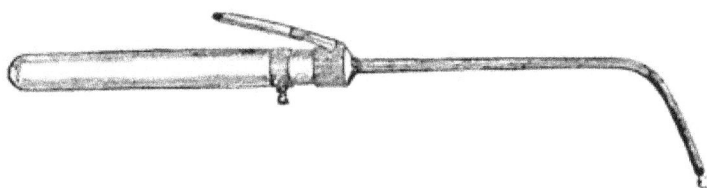

Fig. 6.　Instrument for the introduction of electricity into the larynx.

This instrument is connected to an electric battery by means of a fine wire which is attached to the little ring on the under surface of the handle.　When inserted into the air tube, the spring located on the upper surface of the handle is depressed, which allows the electricity to travel through the instrument and into the larynx.　By this method of treatment the electric current is brought into direct contact with the paralyzed cords.　This is a great improvement upon the old method of application, which simply consisted in the electrization of the outside of the throat through the agency of sponges.　I have frequently seen cases of paralysis of the vocal cords subside after a few direct applications of the laryngeal electrode when persistent treatment externally with the electric sponges have failed to give any relief whatever.

Dr. WHITFIELD WARD.

208 West 25th Street.